Shell Seeker

The Life, Work and Adventures of a Blind Biologist

NSTA Press® and NSTA Kids®
Cathy Iammartino, Chief Product Officer
Emily Brady, Author & Publishing Services Director
Tia Embke, Product Management Director

Design, Production, And Project Management
KTD+ Education Group

National Science Teaching Association
Bob Lay, Interim Chief Executive Officer

405 E. Laburnum Ave., Ste. 3, Richmond, VA 23222
NSTA.org/bookstore
For customer service inquiries, please call 800-277-5300.

Printed in Canada.

28 27 26 25 4 3 2 1

ISBN 9798899770104

A catalog record of this book is available from the Library of Congress.

Shell Seeker

The Life, Work and Adventures of a Blind Biologist

Suzanne Sherman

Illustrated by Linda Olliver

On a small gravel road near the town of Gouda, in the Netherlands, a young boy named Geerat clung to his mother's back as they rode toward the sea. Behind him, his brother Arie bounced along with their father.

The sweet scents of chamomile flowers and fresh grasses wafted in the air. Church bells chimed in distant villages. It was a long ride, but once Geerat heard the roar of the ocean, he knew the shore was near.

Geerat's excitement grew as he smelled the damp, salty air. A vast expanse of sand lay just beyond the dunes.

Geerat's full name is
Geerat Vermeij
(GEER-at Ver-MAY).

At the beach, Geerat and Arie scurried about. They set to work exploring and collecting nature's treasures. Geerat was awed by the details and beauty of seashells. He placed the best **specimens** into his bag to sort at home.

Geerat couldn't see the shells. He turned them over in his hands and ran his fingers along their edges to learn their shapes. When he was three years old, Geerat lost his eyesight to a disease called glaucoma.

Geerat was lucky. Many parents would not let a blind child roam freely on a beach and in the wilderness. But his parents loved nature. They encouraged him to touch everything he encountered. What he couldn't feel, they described to him.

Geerat's youngest years were spent in and out of the hospital.

Back in his cozy home, Geerat emptied his haul to inspect his findings. He sorted the shells by their shapes and textures. Home was his happy place. The smell of homemade applesauce and the sound of sizzling meat drifted from the kitchen. The classical music of Johann Sebastian Bach played on the radio.

Geerat noticed that the music repeated itself. One part sounded almost the same as the next, but slightly different. He observed that his series of shells was similar. The same shapes seemed to appear over and over again. But each shell had **variations**.

He wondered why.

At a young age, Geerat was already thinking like a scientist. He observed nature closely. He asked questions. He was on the path to becoming a great scientist!

Life was not always easy for Geerat. When he was three years old, he went away to a school for blind children. Geerat lived at the school and came home only on holidays and every other weekend. The students at his school were not allowed to own toys or collections. That was a way to prevent the children from fighting.

But Geerat was not happy with this rule. He found interesting objects in the schoolyard and stuffed them into his pockets. When the teacher found Geerat exploring behind bushes, he was scolded sternly.

Geerat felt lonely and trapped.

But Geerat always had a loving family to go home to. As he learned to read and write in braille, so did his family. They surprised him with hand-copied braille books when he came home to visit.

Braille is a system of writing used by people who have low or no vision. Braille uses patterns of tiny bumps that people can read with their fingers.

When Geerat was nine, his family moved to New Jersey, in the United States. He was enrolled in a public school. This was a tough transition for Geerat, but he had a wonderful fourth-grade teacher named Mrs. Colberg.

Mrs. Colberg decorated her classroom with shells she found on her trips to Florida. She let Geerat inspect the collection. Mrs. Colberg's shells were not like any he had found in the Netherlands. As he carefully ran his fingers over each shell, he observed their fancy shapes.

He likened the shells to works of art. Geerat brought in shells from the Netherlands for comparison.

Geerat's enthusiasm for the shells spread. Soon, the classroom collection grew as Geerat and his classmates brought in more shells of their own.

Geerat yearned to learn more about shells and the animals that made them. He had many ways to gain knowledge.

1. He had his family. They read to him and made maps and images he could feel.
2. He used books and magazines.
3. He went to the American Museum of Natural History.
4. He used direct observation.

The more Geerat observed, the more questions he had.

What animals make shells?

Why do they make them?

How do they make them?

Why do shells come in basic shapes?

Why are there so many differences among the shells?

One question stuck with Geerat.

Why are the shells from the Netherlands so plain,

while the shells from Florida are so fancy?

Mollusks

Octopus, Octopus vulgaris
Giant Clam, Tridacna spp.
Softshell clam, Mya arenaria
Carribean reef squid, Sepioteuthis sepioidea
Scaphopod, Dentalium majorinum
Banana slug, Ariolimax columbianus
Deadly cone snail, Conus geographus
Spiny murex, Poirieria zelandica
Lined chiton, Tonicella lineata
Croatian cave snail, Zospeum tholussum
Blue Sea Slug
Glaucus atlanticus
Horse conch, Triplofusus giganteus
Sea Scallop, Argopecten irradians

Geerat learned that animals called **mollusks** make shells. Mollusks are soft-bodied animals. Unlike fish, amphibians, reptiles, birds, and mammals, mollusks don't have backbones.

Many mollusks have shells to protect their squishy bodies. The shells not only provide shelter from the environment, but they also protect mollusks from being eaten. Soft bodies are good eating for many **predators**.

Some mollusks don't have shells. They avoid predators in other ways. Some, like octopuses and squid, move in fast bursts to avoid attack. Others, like slugs, have a slimy coating that makes them difficult to grasp.

Mollusk types are wildly different from each other. They are highly **biodiverse**.

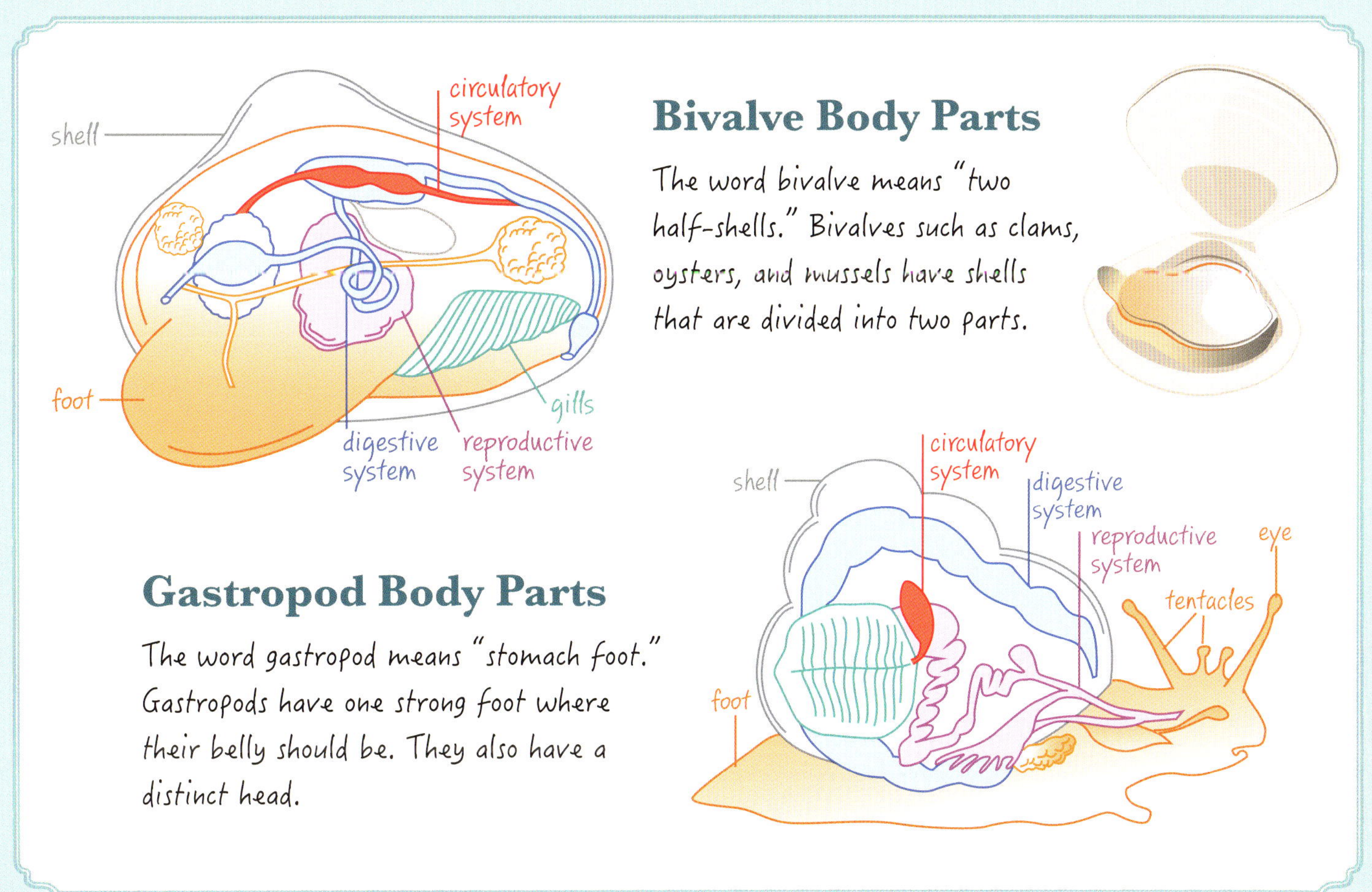

Bivalve Body Parts

The word bivalve means "two half-shells." Bivalves such as clams, oysters, and mussels have shells that are divided into two parts.

Gastropod Body Parts

The word gastropod means "stomach foot." Gastropods have one strong foot where their belly should be. They also have a distinct head.

The nautilus builds larger and larger chambers to live in as it grows.

Mollusks build their shells as they grow. As their soft bodies gain food and get larger, they **secrete** a material near the opening of the shell. This material contains the mineral calcium carbonate. It becomes hard and strong.

In a bivalve, the shell is of two pieces, one on each side of the animal. In a gastropod, the shell is of one piece. The shell forms a **spiral** shape as the animal grows larger.

Spiral growth is a simple theme, yet it can produce many different forms. Geerat was amazed by the biodiversity of gastropods. “It’s like listening to Bach!” he declared.

In college, Geerat studied variations in blue mussels.

Geerat worked hard and earned his way into college. He went to the prestigious Princeton University. There, he learned much more about living things and Earth. He also learned what science was all about. It is not just about finding answers. It is about finding questions. Geerat excelled at finding questions.

How Science Works

In science, observations lead to questions. Questions lead to organized investigations. Investigations lead to more observations, which lead to more questions. This repeated process is called **inquiry**.

Geerat was the only blind student at his college. Laptop computers and screen readers had not been invented. Geerat took all his notes using a slate and stylus. A slate is a flat guide that holds a piece of paper. The stylus is a pointy tool used to make braille words on the paper.

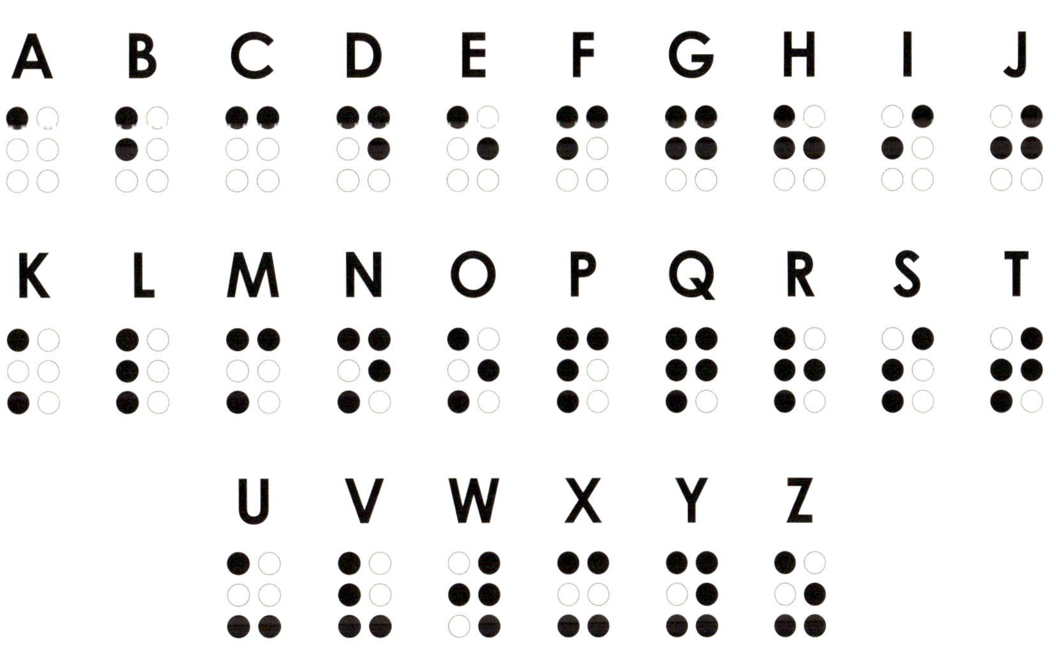

Braille Alphabet

Geerat picked up a shell and noticed the top was broken off. He almost put it back. Then he thought of more questions.

What can we learn from broken shells?

Geerat's inquiry led him on many adventures. When he was still in college, he had the opportunity to visit the **Tropics**. Many people thought Geerat should not go. The work was far too dangerous for a blind person. What if he stumbled upon a deadly snake or fish? What if he were trapped in a terrible storm? What if he reached crumbling cliffs or slippery rocks? But Geerat did go, many times. He encountered all the hazards. And he stayed safe by taking the same precautions that any field scientist should take.

In Costa Rica, the rainforest was a feast for Geerat's senses. He loved the continuous symphony of monkeys, birds, and insects. Questions filled Geerat's mind as he observed the many exotic shells.

What are shells trying to tell us?

What do they say about how the animal lived?

Fieldwork Safety Precautions

- Never go alone.
- Wear proper clothing and gear.
- Move cautiously.
- Know the local dangers.
- Stay alert.

In the Netherlands, the water is cold. Shells don't grow easily in cold water. Animals need energy for many things. They need it to move, grow, and heal. There aren't many predators in colder water, so making thick shells is a waste of energy in that environment. Mollusks with thin, plain shells survive well here.

But what works in one environment might not work in another. In tropical environments, such as Florida and Costa Rica, the water is warmer. Shells form more easily in this water. The biodiversity of predators is high. Mollusks need thick, fancy shells to protect themselves.

Geerat found answers in the patterns he observed. He remembered the question he had in Mrs. Colberg's fourth-grade classroom. Why are some shells so plain and others so fancy?

Now he could finally explain it!

The shells are different because the animals that made them **adapted** to different **environments**.

Geerat's questions didn't stop there. Now that he knew *why* animals had different shells, he wanted to know *how*.

How do animals' bodies come to suit their environments?

Geerat was eager to join the many scientists who study how animals change over time. He wanted to observe shell **fossils**. He would compare them to living animals.

Geerat applied to Yale University to continue his studies. But to get into the school, he had to answer some questions from a professor. The professor was doubtful that Geerat could go further as a scientist. How could he read scientific papers? They weren't printed in braille.

Geerat had to show the professor that he was an expert. His blindness had never stopped him from being a good scientist. Instead, it helped him notice details that others missed.

The professor handed Geerat a specimen from his collection. “Here is something,” he said. “Do you know what it is?” Geerat’s fingers and mind raced. “It’s a *Harpa*,” he said. “It must be a *Harpa major.*” Geerat was right!

The professor was speechless.

“How about this one?” asked the professor, as he tested Geerat again and again. Each time, Geerat correctly identified the species. With this show of skill, Geerat was admitted to Yale.

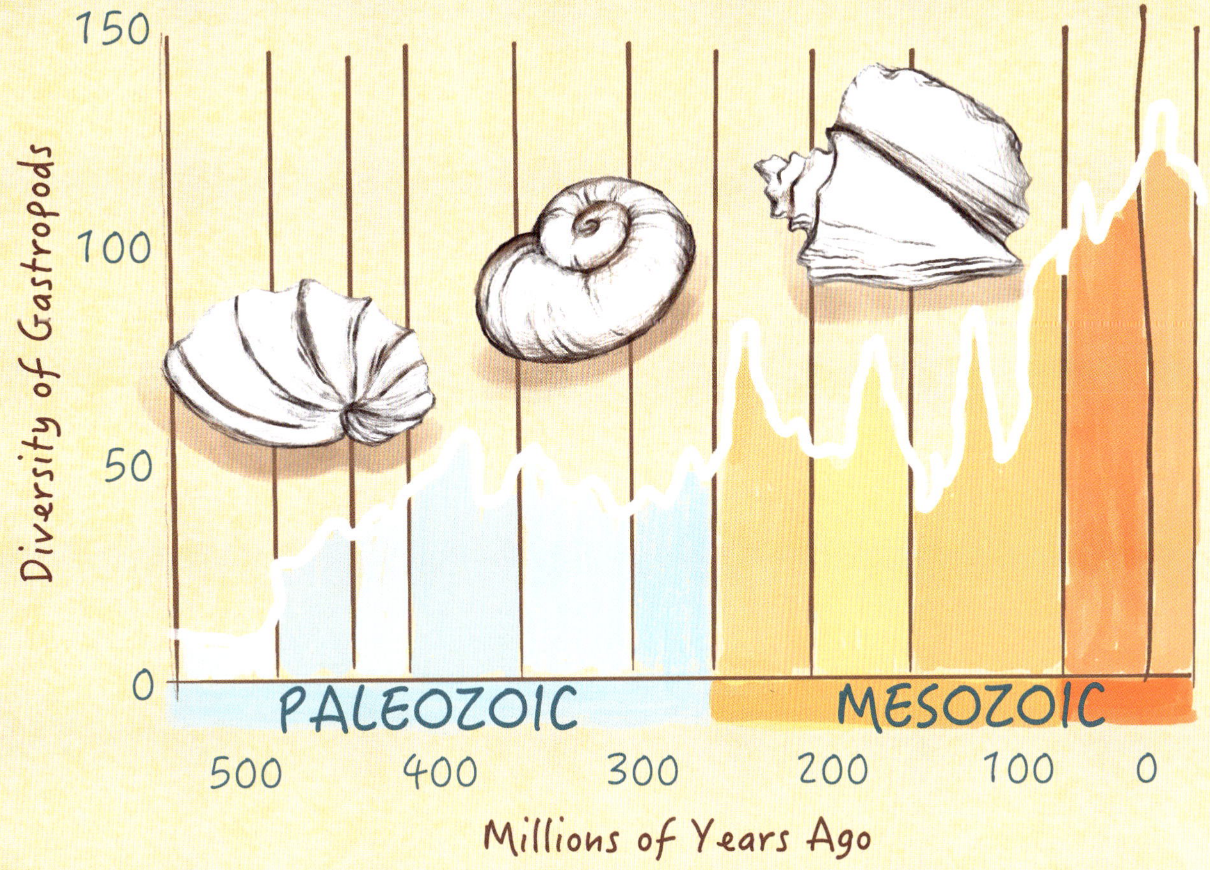

Geerat studied fossils to learn how gastropods became more diverse over time.

It was a good thing Geerat stuck with it. He made one of his greatest discoveries later alongside his wife, Edith. Geerat remembered the broken shells he had found. He knew some of them had been crushed by hungry crabs. He used his imagination.

What can we learn from shell-crushing crabs?

Geerat and Edith went back to the Tropics to learn more. They collected and measured crabs and their **prey**. They compared fossils to living animals. They found some crabs with enormous right claws. Eventually, they had enough **data** to explain how extreme body parts can form—things like massive claws and thick, spiky shells.

The gastropod and its predator changed over time, together. Geerat wrote a study to explain this process. It changed the way scientists think. Geerat helped humans better understand living things and how they change over time.

What questions do YOU have about shells?

How a Snail Got a Spiky Shell

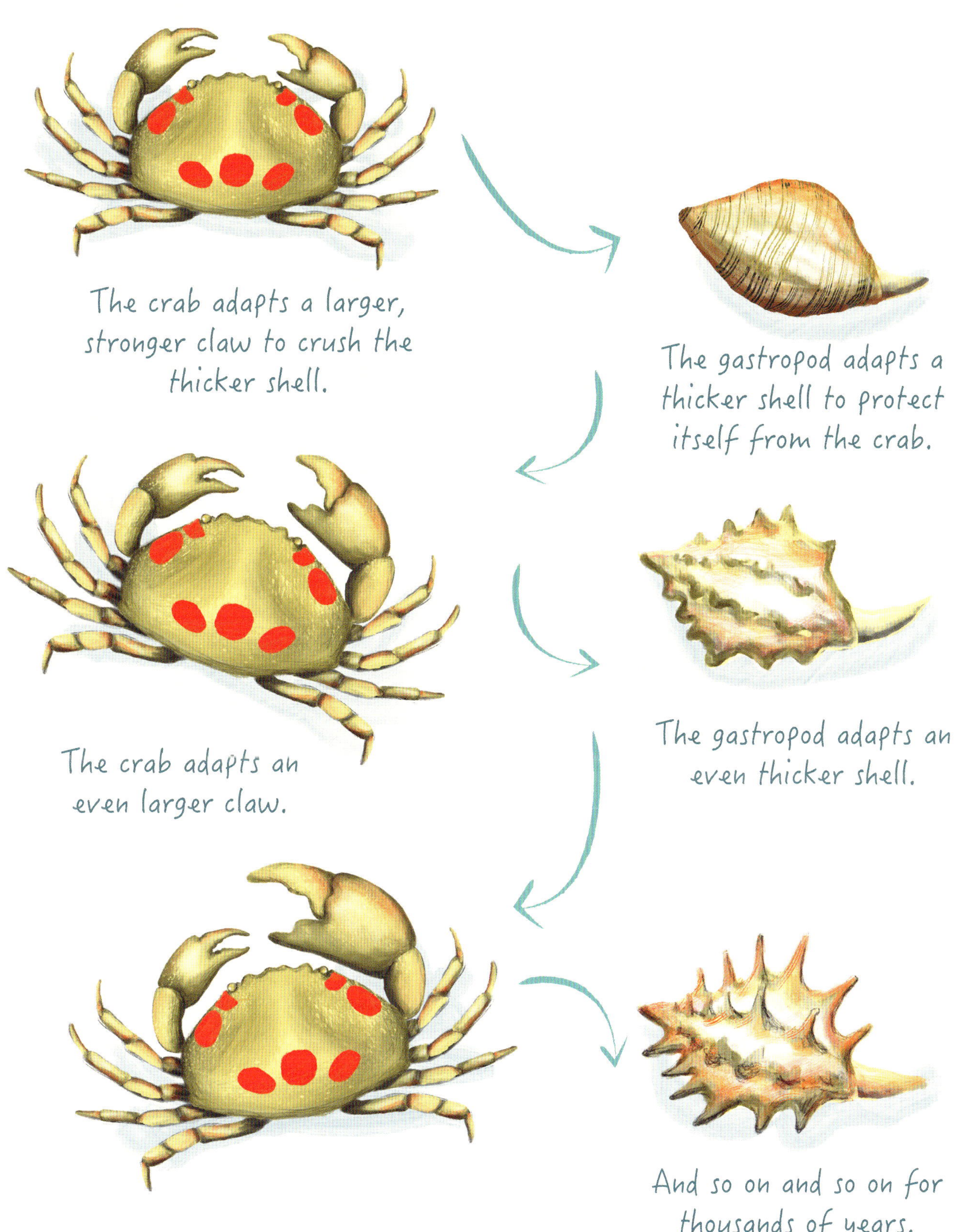

HANDS-ON ACTIVITIES

Analyze Shell Variations

MATERIALS: this book, ruler

1. Observe the blue mussel shells on pages 20 and 21. Describe variations in color and pattern. Describe variations in texture. Describe variations in shape.
2. Use a ruler to measure the length of each shell. Record your data.
3. What patterns do you see among the variations?
4. Choose one variation. Explain how it might help the animal survive.

Write with Braille

MATERIALS: this book, slate and stylus, paper to fit in the slate

1. Refer to the braille alphabet on page 21.
2. Use the slate and stylus to practice writing your name.
3. Write a short message to a partner. Exchange and translate the messages.
4. Describe your experience.

A	B	C	D	E	F	G
⠁	⠃	⠉	⠙	⠑	⠋	⠛
H	I	J	K	L	M	N
⠓	⠊	⠚	⠅	⠇	⠍	⠝
O	P	Q	R	S	T	U
⠕	⠏	⠟	⠗	⠎	⠞	⠥
V	W	X	Y	Z		
⠧	⠺	⠭	⠽	⠵		

Model Internal Structures

MATERIALS: this book, gastropod and bivalve outline handout, yarn of various colors, modeling clay of various colors, scissors

1. Study the diagrams of the internal structures of the bivalve and gastropod on page 17.
2. Use the craft materials to make a model of the internal parts of a gastropod. Include the following: foot, heart, gills or lung, reproductive organ, and digestive system. Do not use glue or tape.
3. Rearrange the materials to make a model of the internal parts of a bivalve. Include the same parts.
4. How do the internal systems compare?

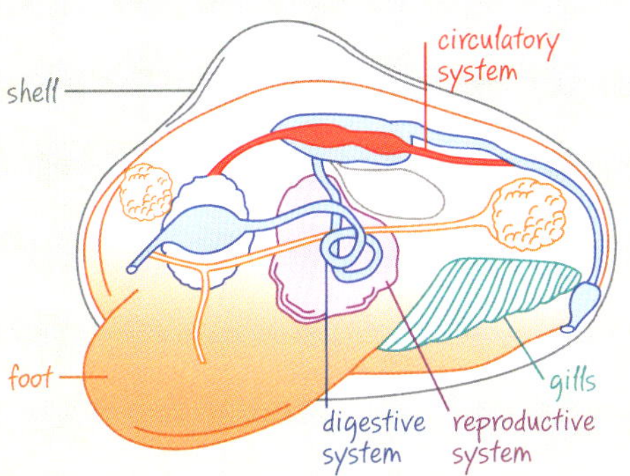

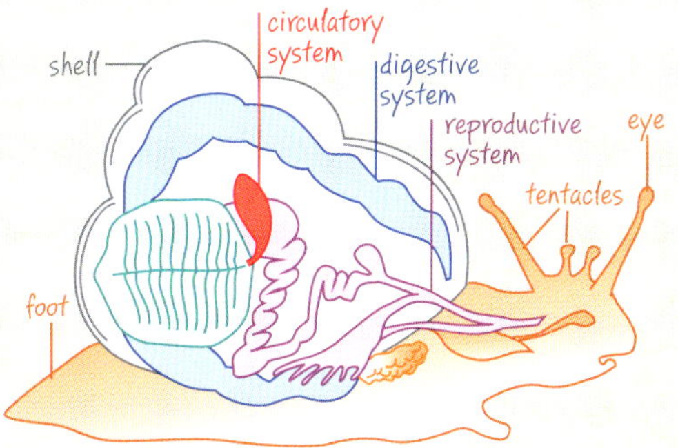

Test Potato Chip Strength

MATERIALS: ridged and smooth clam of similar size and weight, ridged and plain potato chips of similar size and weight, additional materials for testing

1. Compare the textures of the ridged and smooth shells.
2. Plan and carry out an investigation to find out how ridges affect the strength of the shell. Use the potato chips as models.

PROJECT:
Make a Museum Collection

Make and analyze a collection of natural objects. Create your very own museum display to share your findings.

1. Choose a natural object that is made by living things and readily found in your environment. Some examples include shells, leaves, seeds, seed pods, and flowers.
2. Collect many specimens from one type of plant or animal.
3. Carefully observe your specimens. Look for variations in shapes, colors, textures, and sizes. Use tools such as a ruler and a hand lens. Close your eyes to practice better using your sense of touch. Record your observations.
4. Describe the variation in traits. What patterns do you see?
5. Make a graph of one or more traits.
6. Explain which traits likely happened during the plant or animal's life.
7. Explain which traits are likely to be inherited.
8. Do some research. How do the traits you identified help the plant or animal survive?
9. Would the plant or animal survive well in another environment? Explain.
10. Make a museum-type display to share your collection and your findings.

HOW TO
IDENTIFY
OAK TREES
BY ACCORNS

GLOSSARY

Adapt Change to better survive in an environment. An adaptation is a characteristic that a living thing has developed over many generations to suit the environment.

Biodiversity The variety of living things.

Bivalve A type of mollusk that has two matching shells attached by a hinge.

Data Measurements or observations of nature that can be studied for patterns. An investigation can include as few as thirty or as many as thousands of data points. Computers can combine data sets to find patterns among millions of points of data.

Environment The factors in a location that affect the plants and animals that live there. Environmental factors include temperature, water flow, sunshine, food, predators, and competitors.

Fossil The remains or traces of a plant or animal that lived long ago.

Gastropod A type of mollusk that has a muscular foot and a head with tentacles, eyes, and a mouth. Many gastropods have shells shaped like a spiral. Slugs are gastropods that have no shells.

Inquiry	A method of gaining knowledge. Inquiry is driven by observation, question, and organized investigation.
Mollusk	A soft-bodied animal that doesn't have a backbone.
Observation	Using the senses to gather information in a careful, organized way.
Predator	An animal that eats other animals for food.
Prey	An animal that is eaten by other animals.
Secrete	To form and release a substance.
Specimen	An individual natural object or sample to be used for study.
Spiral	A curve that goes around a central point, getting wider or longer (or both) as it continues.
Tropics	Regions of Earth near the Equator.
Variation	The differences between individuals of the same type of plant or animal.

Dr. Vermeij is now a distinguished professor at the University of California. He and his wife have a daughter named Hermine. Dr. Vermeij has written five books and over 300 scientific studies. He is still asking big questions today.

Edith Vermeij and Geerat Vermeij

Hermine Vermeij

Educator Guides

Free educator guides for Grades 1, 2, 3, and 4 are available for download on the NSTA website. Each guide includes additional activities and connections to relevant Next Generation Science Standards and the associated Common Core ELA, Literacy, and Math standards.

ABOUT THE AUTHOR

Suzanne Sherman has always loved exploring the wonders and workings of nature, whether it was building pulley systems for her dolls, observing ant behavior, or exploring the woods and beaches near her childhood home in Washington state. She began her career in neuroscience research and moved on to museum education at Chicago's Museum of Science and Industry. Now, she shares her curiosity as an author of children's science books, bringing the wonders of the natural world to young readers through engaging stories and accurate scientific explanations. Suzanne has written nonfiction readers with publishers including *National Geographic* and Dorling Kindersley, covering topics from electricity to symbiosis to the Milky Way. She lives in La Grange, Illinois, with her husband and two children. You can learn more about her freelance services at treeforteditorial.com and her authoring work at suzanneshermanauthor.com.

ABOUT THE ILLUSTRATOR

Linda Olliver is an award-winning illustrator with thirty years of experience bringing projects to life through imaginative works, both digital and hand-painted. After studying at Parsons School of Design, Linda began her career in advertising before transitioning to editorial publications and marketing. Growing up on the New Jersey Shore, Linda developed a sharp sense of observation through her continual exploration of local rivers and the ocean. She now finds endless fascination in her Baltimore home, with a unique talent for uncovering character and story in her everyday surroundings. While creating infographics may seem different from illustrating children's books, Linda finds great fulfillment in the intellectual challenge of visually communicating complex information.